QUANTUM THEORY

MADE SIMPLE

DISCOVER HOW QUANTUM MECHANICS INTERSECT WITH YOUR REALITY

THEODORE GIESSELMANN

Furthermore, the transmission, duplication, or reproduction of any of the following work including specific information will be considered an illegal act irrespective of if it is done electronically or in print.

This extends to creating a secondary or tertiary copy of the work or a recorded copy and is only allowed with the express written consent from the Publisher.

ALL ADDICTIONAL RIGHT RESERVED.

The information in the following pages is broadly considered a truthful and accurate account of facts and as such, any inattention, use, or misuse of the information in question by the reader will render any resulting actions solely under their purview.

There are no scenarios in which the publisher or the original author of this work can be in any fashion deemed liable for any hardship or damages that may befall them after undertaking information described herein.

Additionally, the information in the following pages is intended only for informational purposes and should thus be thought of as universal.

As befitting its nature, it is presented without assurance regarding its prolonged validity or interim quality.

Trademarks that are mentioned are done without written consent and can in no way be considered an endorsement from the trademark holder.

TABLE OF CONTENT

Introduction

The size of an electron is to a dust speck as the dust speck is to the entire earth" _ Robert Jastrow

James Clerk Maxwell, a great man who brought two halves that seemed irreconcilable to a whole in the 1860s.
His great efforts married electricity with magnetism and revolutionized our understanding of light.
So, we will briefly explore the connections between light, color, and heat. We will encounter a curious physics mystery of the 19th-century and the first step to fall on the road from classical to quantum physics.

Newton's Light Corpuscles

Have you ever imagined of becoming a celebrity in your job or sport? Have you ever thought you could become a champion in both roles?
Sir Isaac Newton was a myth in 2 fields. Not only he invented the laws of motion, but he also laid the foundations for geometric optics.
Newton believed that light was made up of tiny particles that travel in straight lines called rays. These small corpuscles bounce off the mirrors, and when they meet a lens, they always bend slightly at the entrance and exit but travel in straight lines along the way.
An alternative theory of light emerged. Rather than small particles traveling in straight lines through

space, this new theory claimed that light was made up of small vibrations in some underlying medium, like water waves traveling from the wake of a boat to the shore.

Newton did not accept this wave concept because it seemed inconsistent with his geometric approach of lenses and mirrors.

Given Newton's high reputation among physicists, his rejection of the wave of light theory did not emerge for many decades. Nevertheless, even Newton failed to pass the experimental tests, and in the end – two centuries – his classical theory of light particles had been definitively discarded.

Young's Double-Slit Experiment

Now let us go back and explore the theory of the wave of light in detail. First of all, we need to establish some bases on wave phenomena.

Imagine sitting on a fishing boat on a calm, windless morning. Suddenly a motorboat is speeding past you. You should now notice that your boat starts floating up and down. This happens because the past speedboat sends a wave across the water.

Now, imagine that another motorboat is going on the opposite side at about the same time. You will then notice that when the crests of the two waves arrive simultaneously, your boat floats twice as much. This phenomenon is known as constructive interference.

On the other hand, when one crest of one boat's waves coincides with a depression of the others, your boat does not move at all. This is called destructive interference.

For all other cases, your boat's height will be roughly in the middle, as calculated by adding the heights of the two individual waves.
This method of arithmetic addition of wave heights is known to physicists as overlap. More generally, the interaction between multiple waves is called wave interference.

Moreover, it is unexpected observation of this effect that has finally allowed the wave theory of light to emerge from Newton's shadow.

What does this have to complete with light?
Well, imagine now that you are in a bedroom with no light. Open the door of an illuminated corridor. You would expect a large rectangular shape to light upon the bedroom wall as light passes between the door and its frame.
Nevertheless, what would happen if I closed the door more so that the space between the door and the frame became thin? At the beginning of the 19th century, one of Newton's compatriots, Thomas Young, tried to answer.
He cut a small pair of slits into an otherwise opaque object and placed it in a dark room. He pointed a beam of light through the slits, then looked at what appeared on a screen a few meters away. Instead of seeing two thin stripes appear, a theory predicted by Newton (geometric optics), he noticed a series of stripes aligned along with the screen like a fence. The explanation for this observation conflicted with light particles' theory: light was behaving like a wave.
Young speculated that the initial ray was a wave that crossed the room. When interacting with the two slits (points A and B in its original shape), each slit served as the source of a new wave, just like the circular pattern formed when water waves pass through a

narrow channel. As they move away from the cracks, the two waves begin to overlap and interfere. When looking at the darkened screen, the resulting pattern is a series of stripes: dark when a crest of one wave meets a depression of the other (points C, D, E, and F) and light in which the crests of the two waves coincide.

Diffraction patterns can be observed whenever two waves interfere with each other, be they water waves or light passing through a narrow slit in a dark room. Although initially treated with skepticism, Young's work gradually gained general approval.
Before long, he had managed to reverse Newton's particle theory on the light.
The final blow was placed about half a century by Scottish physicist James Clerk Maxwell.

The Birth and Foundations Of Quantum Mechanics

Understanding the Quantum Nature Of Light

physics is trying to interpret the laws as they relate to movement and matter. However, quantum physics is attempting to understand the behavior of the smallest particles and how they move. Such particles contain things like electrons, protons, and neutrons.

A.   Quantum Physics. In Minute' Detail

In its emphasis on microscopic particles, quantum physics explains the particles that make up the tiny particles. The rules regulating macroscopic structures were wrong in setting precedents for smaller domains since the 20th century. The word "quantum" originates from the Latin term meaning "how much." It is used in physics to refer to the tiny matter and energy units whose action is predicted and observed in quantum physics.

Notably, even conditions that exist strongly and constantly, for instance, space and time, have values. However, they seem to be of the smallest degree.
The atom's quantum model is even more complicated than we've seen before; instead of orbiting the nucleus like stars, electrons orbit in blurred, less defined, or cloud-like formation. Furthermore, the final configurations we have learned from the electron sequence (citing the number of electrons in the outer shell) are generally more like probabilities than hard and fast formation.
We bring this up when addressing the quantum nature of light to define the term quantum physics, so you can understand that its purpose is to show the numerical probability of the electron's place at any available time. Therefore, when the word is combined with "the essence of light," you should have a strong understanding of the general working principle.

B. SINGULAR FOR QUANTUM PHYSICS

The possibility that considering anything can conceivably influence the physical cycles that happen is one of a kind to quantum material science. For instance, in what is perceived as wave-molecule duality, light waves doing indistinguishable particles, and these sections additionally show like waves. Put another way, light has the two particles and waves' qualities, and either clarification can depict the activity of light.

In quantum tunneling, the issue can move to start with one area then onto the next without traveling through the mutual space. This offers a route to an advanced application where data can immediately go over great separations. Through quantum material science, we find that a great deal of the universe can be spoken to as a progression of probabilities.

There is a wide range of fields of quantum material science. The one that shines, especially on the conduct of lights (photons), is known as Quantum Optics. By investigating Quantum Optics, you will find that the development examples of individual photons (light shafts) legitimately influence the warm light.

The serious and flexible instrument known as the LASER is only one of the numerous basic side-effects of

This contrasts with the more common study of light, Classical Optics, acquired by Sir Isaac Newton, where the light was spotted as though it had lone particle belongings, meaning that it journeyed in a straight line,

returned from objects with which it emanated into contact and dispatched through objects with minimum resistance.

C. PHOTONS

To more readily comprehend what is suggested when the term photon is utilized, let us direct our concentration toward the Photon Theory of Light. In this specific sense, a photon is a watchful group (or quantum) of electromagnetic (or light) vitality.
Existing in a vacuum and steady movement, photons have a consistent speed of light for all eyewitnesses. It occurs at the vacuum speed of light (more by and large alluded to as the speed of light), which is utilized.
$C = 2.998 \times 108$ m/s long.
Fixated on the Photon Theory of Light, the principle attributes of photons are as per the following:
*	They prop up at a consistent speed, $c = 2.9979 \times 108$ m/s (light speed)
in free space.
*	They are known to have zero mass and zero rest vitality.
*	They convey vitality and energy, which relate with the recurrence nu and frequency lambda, (and p, force) of the electromagnetic wave by E
$= hv$ for and $p = h/lambda$
*	They can be destroyed or created when radiation is absorbed or emitted.
*	They have the ability to have particle-like interactions, for example,
electron impacts and other moment pieces.

D. Quantum Optics. Basic Understanding

To comprehend the quantum properties of light better, it may be useful to apply a portion of the related cycles (retention, emanation, and invigorated outflow) to the Laser, as this is one of the most notable utilization of quantum optics. Generally, these equivalent three attributes might be summed up to other light sources in differing degrees.

Electronic advances are normally the kinds of advances that transmit or ingest noticeable light. Simply envision an electron moving between measured nuclear vitality levels to perceive how this function.

For the Laser to work productively, the invigorated outflow of light is significant. Animated light emanation is utilized to give the enhancement needed to perform imaging work appropriately.

The exceptional property known as cognizance is the consequence of the animated emanation measure.

The normal upgrade triggers the emanation occasions that are liable to give the enhanced light. These partners the released photons in the ideal advance arrangement where every photon has the last stage relationship.

This type of intelligibility (relative arrangement) is characterized in two different terms: worldly cognizance and spatial soundness. Both end up being extremely critical in the advancement of obstruction used to produce 3D images.

Note: Ordinary light isn't intelligent because it starts from autonomous iotas that transmit around 10-8 seconds in time scales.

Although there might be some level of rationality in chose sources, for example, the mercury green line and other valuable ghostly sources' sprinkling, their consistency is not about what is contained in the Laser.

FEW SPECIFIC CHARACTERISTICS THAT ARE SINGULAR TO LASER

Light Include:

1.   Cognizance: This is the property where different laser shaft parts are identified with one another in a stage relationship. At the point when kept up over a long enough period, impedance impacts can be seen or photographically recorded. Intelligence is the factor that makes the possibility of visualizations conceivable.

2.   Monochromatic: comprising of one frequency, the laser light begins from the invigorated outflow of a solitary scope of nuclear vitality levels.

3.   Collimated: Since they need to go through the mirrors a few times at amazingly opposite edges, the ways influenced by the intensification are known for their capacity to rebound back between the laser hole's reflected finishes firmly. Of this reason, the laser radiates have been intended to be extremely tight and restricted in their capacity to expand.

E.   Photon and Probability

There are two manners by which the likelihood can be applied to photons' activity: the likelihood can be utilized to quantify the conceivable number of photons in a specific state, or the likelihood can be utilized to gauge the opportunity that a solitary photon will be in a specific state.

Since the previous understanding is in opposition to Newton's Energy Conservation Principle, the last translation is the most practical other option.

Following crafted by physicist Thomas Young during the 1800s with a twofold cut investigation, Paul Dirac (1902-1984) was a hypothetical British physicist and one of the key pioneers of quantum material science, talks about this standard in his refreshed variant:

Indeed, even before finding quantum material science, individuals realized that a connection between light waves and photons must have a real character. Notwithstanding, it was not so much sure that the wave work gave data on the likelihood of a photon in a fixed situation rather than the absolute plausible number of photons in that area.

This is an important distinction and can be clarified in the following way. Assume we have a light emission comprised of numerous photons isolated into two pieces of equivalent force. If the bar is connected to a sensible measure of photons, half of the likely number of photons thought to be reasonable to every distribution. At the point when the parts are made to meddle with one another, one photon in one segment ought to upset the other.

According to the old theory, the two photons would either must cancel out one another, or they would have to generate four photons. Any outcome would contradict the principle of energy conservation. Thus, according to the new theory, because the photon only slightly affects both two elements, the question of relating the wave function to the probabilities for one photon only becomes a non-issue. Each photon can only cause interference with itself in this system, preventing two photons' potential occurrences.

The Finding of The Scientifically Destroyed Materialism

The quantum model of the universe is an endeavor to liberate the huge explosion from its creationist suggestions. The advocates of this model depend on perceptions of quantum material science (subatomic material science). In quantum material science, it very well may be seen that subatomic particles show up and vanish precipitously in a vacuum. On the off chance that a few physicists decipher this perception so the matter can emerge on a quantum level, it is a property that alludes to issue. A few physicists attempt to clarify the rise of issues from non-presence during the production of the universe as a property identified with an issue and to speak to it as a feature of customary laws.

Nonetheless, this logic is certainly impossible and can not the slightest bit clarify how the universe was conceived. William Lane Craig, creator of The Big Bang: Theism and Atheism, clarifies why:

A quantum age material of the mechanical vacuum is a long way from the standard thought of a 'vacuum,' amounting to nothing. This is 'nothing,' and along these lines, material particles don't emerge from anything.

Matter doesn't exist in initial quantum material science. What happened is that ecological vitality abruptly gets matter, and afterward, similarly as out of nowhere, it becomes vitality once more. So, there is no existential condition from nothing, as it is guaranteed.

As indicated by Isaac Newton, the light was the development of a substance known as the body. The reason for ordinary Newtonian material science - which was recognized until the divulgence of quantum physical science - was that light included absolutely of a grouping of particles. In any case, James Clerk Maxwell, a nineteenth-century physicist, recommended that light displays wave development. Quantum speculation has obliged this most basic conversation in material science.

In 1905, Albert Einstein ensured that light had quantum or little bundles of essentialness.
These imperativeness packages were called photons. But depicted as particles, photons have been believed to carry on in the wave development proposed by Maxwell in 1860. Accordingly, the light was an advancement wander among waves and atoms (George Gilder), an articulation that showed a great irregularity to the extent of Newtonian material science.

Following Einstein, the German physicist Max Planck dissected the light and astonished the whole consistent world by finding that it was both a wave and an atom.
As shown by this thought, which he proposed under the name of the quantum theory, the vitality was passed on as deterred and discrete packs, instead of being straight and reliable.

In a quantum event, the light exhibited both particle and wave-like properties. A wave in space went with the particle known as a photon. The light going through space like a wave yet went about as a working particle when it encountered a tangle.

To put it out of the blue, it showed up as essential until it encountered an obstacle. By at that point, it showed up as particles like they were contained minute material bodies reminiscent of grains of sand.

Amit Goswami Says This Of The Discovery About The Nature Of Light:

Right when light is viewed as a wave, it appears to be prepared to wind up in (at any rate two) positions at the same time, like it experienced the aperture of an umbrella, making a diffraction plan. In any case, when we get it on photographic film, it shows up sagaciously point by point like an atom shaft. Subsequently, the light ought to be both a wave and an atom. It is one of the limits of ancient material science and away from language. The chance of objectivity is furthermore being referred to; does light or what light is depending upon how we see it?

Specialists now not, at this point, acknowledged that the issue was included inorganic and self-assertive particles. Quantum physical science had no materialistic noteworthiness because there were unnecessary things like an issue.

De Broglie's revelation was remarkable; in his investigation, he saw that even subatomic particles showed wave-like properties. Indeed, even particles like the electron and the proton had frequencies. Figuratively speaking, despite the materialistic conviction, there were surges of unimportant essentialness inside the atom, which authenticity called a fundamental issue. Much equivalent to light, these little particles in the particle a portion of the time acted like waves and showed the properties of the particles in others. Contrary to common wants, the fundamental issue in the molecule could be recognized at explicit events, yet evaporate on various occasions.

This essential exposure showed that what we imagine as this current the truth was a shadow. The issue had moved through and through away from the field of material science and was moving towards otherworldliness.

Physicist Richard Feynman portrayed this entrancing reality about subatomic particles and light:

by and by, we know how electrons and light act. However, what might I have the option to call it? Right when I state they act like particles, I set up an improper association. Whether or not I express that they show like waves, this isn't right. They work in their preeminent way, which could be called quantum mechanics. A particle doesn't act like a weight that hangs and swings on a spring. It disdains a littler than a typical depiction of the close by planetary gathering with little planets moving around and around. No more does it appear like a cloud or haze, including the middle. It takes after nothing you've any time seen already.

There is, regardless, one unraveling. In such a way, electrons continue basically like photons. They are both crazy.

How they act requires a vast amount of imaginative personalities to esteem them since we will all delineate something different from all that we know. Nobody knows how it is. It gets from the workplace of the pioneer of the Bohr gathering, who suggested that the physical reality proposed by the quantum speculation is the information we have about a system and the examinations we make subject to this information. In his view, these assumptions made in our psyches had nothing to do with external reality. To put it plainly, our inward world had nothing to do with the real, external world, which was the essential enthusiasm of Aristotle's physicists to the current day. Physicists have abandoned their old thoughts regarding this view and have concurred that quantum understanding is just our insight into the physical framework.

The material world that we can see exists just like data in our minds. As it were, we can never have direct encounters with the issue in the rest of the world.

Jeffrey M. Schwartz, neuroscientist and educator of psychiatry at the University of California, portrayed this end from the Copenhagen translation:

John Archibald spoke: 'No marvel is a wonder until it is a watched wonder.'

AMIT GOSWAMI EXPANDED THIS RESULT:

Assume we ask; is the moon there if we don't look at it? To the extent that the moon is, finally, a quantum object (made altogether of quantum objects), we need to state no, says physicist David Mermin.
Maybe the most significant and slippery presumption that we consider in youth is that of the material universe of articles out there, paying little heed to who the eyewitnesses are. There is proof of this suspicion. For instance, if we take a gander at the moon, we will discover the moon where we anticipate that it should be along its traditionally determined direction. We venture that the moon is consistently there in space-time, in any event, when we are not looking. Quantum material science says no. If we don't look, the rush of the moon's chance is spreading, but of a limited quantity. At the point when we look, the wave stops in a split second; like this, the wave couldn't be in space-time. It bodes well to receive an otherworldly philosophical hypothesis: There is no item in space-time without a cognizant subject taking a gander at it. Of course, this applies to our world of perception. The presence of the moon is apparent in the outside world. In any case, when we take a look, everything we experience is our view of the moon.
Jeffrey M. Schwartz embedded these lines 'The Mind and the Brain'
relating to the reality showed by quantum material science:
The job of perception in quantum material science can't be exaggerated. In traditional material science [Newtonian physics], watched frameworks have a presence autonomous of the psyche that observes and inspects them. In quantum physics, however, a

physical quantity has real value only through an act of observation.

Schwartz also summarized the opinions of various physicists on this topic:

As Jacob Bronowski wrote in 'The Rise of Man,'

'A goal in the natural sciences was to provide an accurate picture of the material world. One of the achievements of physics in the 20th century was to demonstrate that this goal cannot be achieved.' Heisenberg said that the concept of objective reality was 'so blurry.' In 1958 he wrote that 'the laws of nature that we mathematically formulate in quantum theory no longer deal with the particles themselves, but with our knowledge of elementary particles.' 'It's wrong,' Bohr once said, 'to think that the task of physics is to find out what nature is like. Physics is about what we can voice about nature.'

After the most fascinating and sensitive experiments that the human mind has been able to develop over 80 years, there are now no opinions on quantum physics that have been proven decisively and scientifically. Nor can any objection be raised to the conclusions of the experiments carried out. Scientists have tested quantum theory in hundreds of possible ways and have received the Nobel Prize on several occasions for their work.

Matter, the most basic concept of Newtonian physics and once considered unconditionally as absolute truth, has been removed. The materialists, supporters of the old belief that matter was the only and last block of existence, were confused by the lack of matter suggested by quantum physics. Now you have to explain all the laws of physics in the field of metaphysics.

The shock this caused to materialists in the early 20th century was far greater than can be expressed in

these lines. But quantum physicists Bryce DeWitt and Neill Graham describe it:
'No development of modern science has influenced human thought as profoundly as the advent of quantum theory. Torn from secular thought patterns, the physicists of a generation were forced to face a new metaphysics. The difficulties that caused this reorientation to continue today. Physicists suffered a severe loss; their hold on reality.'

RESISTANCE

In the eyes of electricity, not all materials are created equal. Some materials are quick to allow the flow of charges, while others do their best to hinder electric flow. Commonly, a material's ability to conduct heat is a good indicator of its willingness to work electricity. Metals are, in general, great conductors in terms of heat and electricity. The opposite is true for water, wood, glass, and air. There is a neat phenomenon that can demonstrate this. When two surfaces are in contact, friction can cause a transfer of charges and thus a build-up of potential difference. For example, wool clothing rubbing against the ground commonly results in such an imbalance (especially on cold and dry days). As long as you don't touch anything conductive, the extra charges will simply sit on your clothes and do nothing. The air surrounding you is not conducive enough to get them moving, despite the potential difference. But touch a metal doorknob and - zap - the charges flow, leaving you slightly shocked. How can we define electric resistance mathematically? Suppose you create a particular potential difference within an object. In a material with only a little electric resistance, a relatively strong current will result from that. The same potential difference, however, will only lead to a weak current if the electric resistance is high. So it makes sense to define electric resistance as the ratio of potential difference V to current I:

$$R = V / I$$

At an absolute value of V, a large current will lead to a small amount of resistance, while a small current will do the opposite. This is just what we wanted to reflect. Materials with low electric resistance are called

conductors (metal), and those on the other end of the scale insulators (water, wood, and so on).
Let's go back to the fluid flow analogy. As mentioned, we can interpret voltage as the pressure difference that will cause the flow to occur. The current is then equivalent to the speed of the flow. What does that make electric resistance? In this context, we can interpret it as friction. A high value of resistance does to electric discharge what a rough surface does to fluid flow. It hinders the formation of a strong current (flow speed), given a potential difference (pressure difference).
On a microscopic scale, electric resistance arises in part from collisions within the material. A potential difference spawns an electric field throughout the conductor, which in turn exerts a force on the charges, usually electrons, causing them to accelerate. Their motion comes to an end quickly and abruptly when colliding with an atom that happens to be in their path. Since the electric field is still active, the charged particle will accelerate again and collide with the atom in its way. This is an electric flow on the smallest scale. It is not a continuous and gentle stream of electrons but rather a series of rapid accelerations and violent collisions.

Naturally, the more atoms in the way and the bigger they are, the harder it is for the electrons to flow at a high average velocity (producing a high charge flow rate = current).

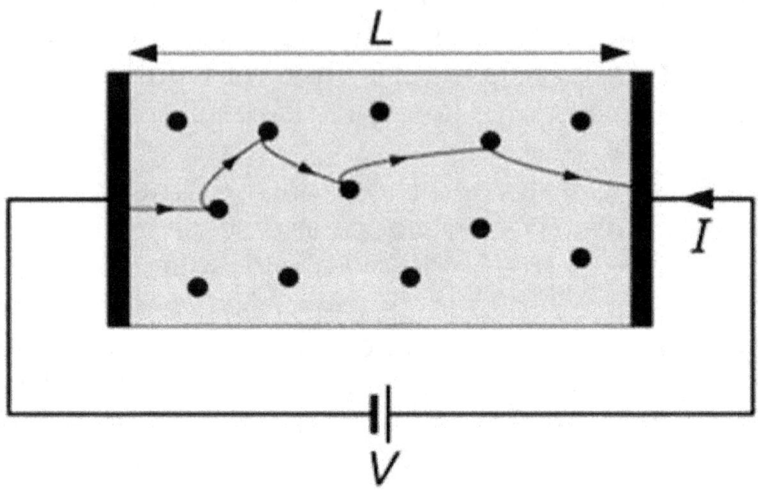

(Visualization of the path an electron takes through a conductor. Not a straight-line, but a series of acceleration and collision)

A word of warning: electric flow can cause severe injuries and even death. If you work with electricity and you are not one-hundred percent sure what you're doing, stop immediately. Even a seemingly harmless
7.5 W holiday light can burn or kill a person if handled carelessly. Never try to fix electric appliances if you are not a professional. About 60 people die in the US each year from just that. If a device develops smoke, gets unusually hot, or smells "burned," turn it off immediately and don't turn it back on until a professional has taken care of the problem.

STEWART-TOLMAN EFFECT

The American physicist's Dale Stewart and Richard C. Tolman thought up. They performed an interesting experiment that is capable of demonstrating the nature of voltage in an easy-to-understand manner. Before we turn to the investigation, imagine a truck that transports a particular gas in a container.
Initially, both the car and the gas are at rest, and as is the case for all bodies in the universe, they tend to stay in the current state of motion. However, when the truck driver steps on the gas pedal, a force results that causes the truck to accelerate.
The gas in the container is not directly affected by force. It remains at rest for the first milliseconds. Only when the back wall of the box starts pushing it forward will it accelerate as well. This leads to the gas accumulating in the back. As long as the truck accelerates, there will thus be a pressure difference within the container. Once the truck assumes a constant velocity, the pressure equalizes.
In an electric conductor, the charges (usual electrons) are present even with a potential difference or current. They sit there, waiting to be moved by the presence of an electric field. If you accelerate the conductor, this electron gas behaves no differently from the gas in the above example. The acceleration causes the electrons to accumulated in the back. If you attach a voltmeter to the conductor at this point, it will show the presence of a potential difference. Once the conductor

transitions into a state of constant velocity, the electric potential quickly disappears.
This so-called Stewart-Tolman effect, the appearance of voltage when accelerating a conductor, shows that

an uneven distribution of charges causes potential difference and that if there's nothing to maintain it, the resulting current will equalize it. The experiment also "demystifies" electricity to some extent by demonstrating that charged particles obey the same mechanical laws as all other matter. When you are uncertain of how to interpret voltage, just think back to this experiment.

Piezoelectricity

In 1880, the brothers Pierre and Jacques Curie made
a curious discovery that would lead to many great
applications over the years to come. They found that
when you deform certain materials (mainly crystals),
a potential difference arises. This intriguing behavior is
called the piezoelectric effect. Shortly after the
discovery, the Nobel laureate Gabriel Lippmann
suggested that it should be possible to reverse this
effect. The Curies quickly showed that this is indeed
the case: when you apply voltage to such materials,
they deform as a result of it.
To explain this behavior, we need to take a look at the
centers of symmetry of the positive and negative
charges in the crystal. In an unstressed state, the
centers coincide, and the crystal appears neutral.
When the crystal is deformed, the centers shift and
move apart. This leads to a net charge and, thus, a
potential difference. In the process, the mechanical
energy that was used to deform the object turns into
electrical power.
The effect can be used for pressure and acceleration
sensors since both will cause a body to deform. One
notable example is the use of piezoelectric sensors in
microphones. They can pick up the pressure
fluctuations caused by sound waves and transform the
sound into the electric flow. The possibility of
reversing the effect allows us to build devices that
turn electric discharge into pressure waves and thus
sound (loudspeakers).

An unusual application of piezoelectricity can be found in a Rotterdam night club with the fitting name "Watt." The owners built the floor on a large number of crystals that compress and expand as the visitors dance the night away. These deformations create enough electricity to power the club for most of the night.

LIGHTNING

What comes to mind when thinking of electricity? For most people, the answer indeed is lightning. Lightning strikes are rare opportunities to observe electric flow in action, something that is usually hidden from us (for good reasons) using insulation and casings. Let's take a quick look at this spectacular natural phenomenon.
On hot days, the sun's rays will heat the ground considerably, which in turn causes the temperature of the air near the bottom to rise. Given the right circumstances, this can lead to convection: parcels of air become buoyant and grow. As they do so, their temperature drops, and once it goes below the dew point, the water vapor within the parcel condenses. A cloud is born.
The formation of a cloud is associated with a separation of charges. Observations show that the top of the cloud ends up positively charged, while at the bottom, negatively charged particles dominate. As of now, no theory exists that can fully explain this electrical structure. Several ideas have been proposed that might provide an answer and are currently being researched. We will not go into the mechanisms for further discussion, and it is sufficient to know that this separation of charges leads to a potential difference

not only within the cloud but also between the cloud and ground.

There is a natural tendency to even out any inhomogeneous distribution of charges via electric flow. This is not possible in the case of a cloud because air acts as an insulator. The charges "want" to flow, but cannot due to the high electric resistance of the surrounding air. However, once the potential difference reaches a critical value (around 50,000 volts per meter), a low-resistance channel of ionized air forms that lets the charges pass. Now the dam is broken, and the resulting current is enormous. Billions of electrons and positive ions race at speeds of 360,000 mph through the channel. This causes the temperature to rise almost instantly to 22,000 K and beyond. The rapid thermal expansion of the lightning channel results in the thunder so typical for a lightning strike. Commonly we think of lightning as going from cloud to ground. However, the opposite is also possible and frequently happens at skyscrapers and in the mountains.

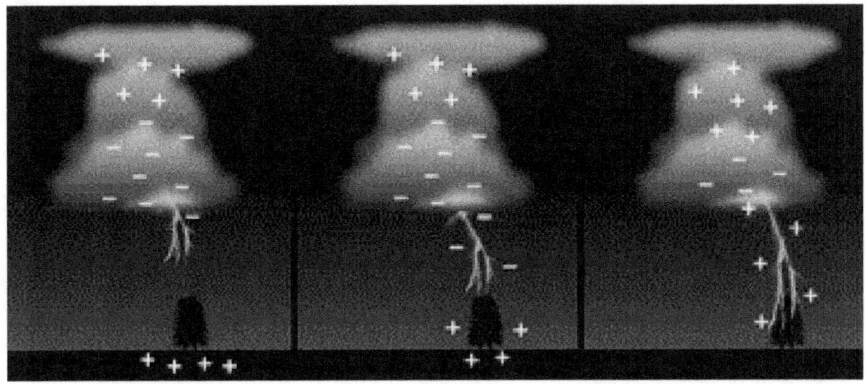

Thunder travels with about 340 m/s, the speed of sound, while the light coming from the electric flow reaches you almost instantly. This provides a simple way to estimate the distance to the thunderstorm. When you spot the flash, start counting the seconds; when you hear the thunder, stop. Divide the result by three - this is the approximate distance in km (divide by five for the distance in miles).
When lightning strikes a person, the electric flow can cause severe injuries and even death. In the US alone, there are, on average, 35 lightning fatalities per year, the majority of males (74 %) between the ages of 20 and 59. If a thunderstorm approaches, go to a safe shelter (building, car). Never seek shelter in sheds, boats, or under trees. Don't take a bath or shower and use your telephone only in case of an emergency. If no shelter is existing, go to a low spot away from trees, fences, and poles. Tingling sensations or electrified hair are signs of immediate danger. In this case: squat and make yourself the smallest target possible.

Holography and Quantum Gravity

The Myth of Gravity

My eldest son reads a primary science reading explaining gravity. He says to me, "Papa, look, every time I jump in the air a force pulls me back down to Earth, do you know why?"

Ironically, I said no. He moved on to explain to me that it is simply because of gravity. Gravity pulls everything down to Earth, and all masses attract each other, he added. Finally, he went on to ask me why gravity behaves in this way. I just said to him, son, this is just the way nature works. Newton, in response

to a similar question, said, "hypothesis non-fingo."
This translates as "I do not have a clue."
Astonishingly, such a simple and straightforward
question about a common phenomenon we experience
every minute of our lives has not found a direct
answer for many centuries, even from the brightest
physicists. Why is gravity there in the primary place?
What is gravity? Commonly, it is defined as an
attractive force that all matter possesses. Matter
attracts other issues, and the strength of the
attraction depends on two things, the mass of the
objects and the distance between them.
Gravity and its subordinate attractive energy originate
from the Latin word gravitas, from gravis, i.e., hefty.
It has been accounted for that the advanced
significance of fascination didn't show up until
Newton's time. For researchers like Galileo and
Copernicus, up to the start of the twentieth century,
the word gravity had no concrete significance. We gain
from old-style material science that attraction or
gravity is a characteristic wonder where articles or
bodies with the mass draw in each other. Attractive
energy is liable for keeping our planet in its circle
around the Sun. It is likewise answerable for
observing the Moon in its rotation around the Earth.
Without gravitation, we would witness chaos in the
universe as it causes dispersed matter to coalesce.
Most important of all, gravity is responsible for the
formation of tides, natural convection, etc. For
instance, we learn that in space, objects maintain
their orbits because of the force of gravity acting upon
them. Despite the successes of the SM of particle
physics in unifying the three fundamental forces of
nature, the fourth—namely gravity—remains elusive.
This is due perhaps to our lack of understanding of its
origin and why it is there in the first place. Among all-
natural forces that exist, gravity is the force we are

most familiar with, but so far, its origin has eluded even the brightest physicists.
Newton's physics explains how an apple falls towards the Earth. Einstein went a step further by explaining gravity as the warping of the fabric of space-time.
 All these theories so far have described how gravity works, but their shortfall remains in their inability to explain how it arises. For instance, Newton had a lot of reservations and doubts about the action of gravity at a distance. For centuries, this question, among others, has not been satisfactorily answered.

How can a massive object attract another at great separation without any mediation?
Another question that has also remained unanswered is how to explain the attraction between two objects without time to action? Recently, many publications have pointed out that gravity might not be a fundamental force at all, but rather an emergent property of the deeper underlying structure of the universe. Since the publication in 2010 of Verlinde's paper entitled "On the Origin of Gravity and the Laws of Newton," many follow-up papers have examined this idea in a new light. The overwhelming feeling that arises from those publications is that a new worldview emphasizing the primary role of information over matter and energy is likely to take center stage.
 However, I would first like to look at the history and implications of the theory of gravity, from ancient Egyptian times to Einstein, as well as recent attempts by several researchers, who believe that the universe could be emergent and holographic. I will also examine how this new model, in which gravity may not be a fundamental force, could change our picture of space-time, giving a boost to the physics of computation and information—what we now call digital physics.

UNIFIED ENERGY AND UNIFIED MATTER

his generates us to the concept of the Unified Field. This also brings us to the Unified Force, Unified Matter, and the Theory of Everything.

In this part, I offer that the "energy string" is the fundamental entity of the universe. These energy strings are the special hybrid of mass and energy from which all the types of particles, and all the forms of energy, are created.

I also believe that the five types of energy strings provide most, if not all, of the properties found in the universe.

Therefore, this leads me to believe the above set of concepts will ultimately move us toward the "Unified Field Theory."

UNIFIED FIELD

Unified field hypothesis, in molecule material science, a push to characterize every primary power and the connections concerning elementary particles in specifications of a singular hypothetical structure. In material science, powers can be named by fields that referee dealings among disengaged things. During the nineteenth century, James Clerk Maxwell passed on the underlying field hypothesis in his way of thinking of electromagnetism. At that point, in the early part of the twentieth century, Albert Einstein set up general relativity, the field hypothesis of attractive energy. Consequently, Einstein and others attempted to create a bound together field hypothesis in which electromagnetism and gravity would surface as various parts of a solitary major field.

They fizzled, and right up 'til today, gravity stays past endeavors at a bound together field hypothesis.

At subatomic separations, fields are portrayed by quantum field hypotheses, which apply the thoughts of quantum mechanics to the whole ground. During the 1940s, quantum electrodynamics (QED), the quantum field hypothesis of electromagnetism, grown completely settled. In QED, charged particles associate as they produce and retain photons (minute parcels of electromagnetic radiation). As a result, trading the photons in a round of subatomic "get."

This hypothesis works so well that it has become the model for speculations of different powers. Moreover, I have said that the "field" is just the streaming of vitality strings. Accordingly, on the off chance that all "fields" are kinds of vitality strings, at that point, we can make a "Brought together Field Theory" in light of "bringing together" these vitality strings.

In particular, I accept that these sorts of vitality strings are varieties of one essential kind of vitality string. I see the kinds of vitality strings recorded and portrayed above as branches off a tree. We know we have comparable branches. We should simply locate the principle part of the tree. At the point when we distinguish that focal part, we will, at that point, have the option to bring together all fields into one complete picture.

UNIFIED MATTER

These energy strings also lead us to Unified Matter. There are many scientists, including Heisenberg, who talk not only of a unified field but a "Unified Matter." Heisenberg believed that the quarks being discovered actually had similar properties, and therefore would lead us to the understanding of a Unified Matter. Indeed, this we can do with our energy strings. The energy strings I have proposed are the hybrid of energy and matter. Therefore, by unifying these energy strings, we will not only be unifying our energy fields but unifying our matter as well.

THE FUTURE OF UNIFIED FIELD-BASED ON ENERGY STRINGS

I do believe that the future of the Unified Field, the Unified Force, and the Unified Matter will all be developed from the five fundamental types of Energy Strings I have listed and mentioned above.
For the first time, we have a common language and a standard structure for each type of field, each type of force, and all the observed properties in any particle. For the first time, we have commonalities, where every field, force, and particle can begin to look like the other fields, forces, and particles. This is a big step.
Now we can take these commonalities and find the True One Unifying Concept behind them all. Indeed, from just One Universal Energy, we can create all things. All energies, all matter, all motions, and all observations can be explained based on these few concepts.

These secrets have now been discovered. They will gradually be revealed, through a future series of publications, leading ultimately to the Unified Energy Solution and the Theory of Everything.

SUPERCONDUCTORS

NO RESISTANCE

Then resistance disappears? The answer to this question was Kamerlingh Onnes as early as 1914. He proposed a very ingenious method of measuring the resistance. The experimental scheme looks quite simple. Lead wires from the coil in a cryostat omitted - an apparatus for carrying out experiments at low temperatures. The cooled helium coil is superconducting. In this case, the present flowing through the coil, creating a magnetic field around it, which can be easily detected by the deviation of the magnetic needle located outside the cryostat. Then close the key, so that now the superconducting stroke was short-circuited. A compass needle, however, was diverted, indicating the presence of current in the coil is already disconnected from a current source. Watching the arrow for a few hours (until evaporating the helium from the vessel), Onnes had not noticed the slightest change in the deflection direction. According to the results of the experiment, Onnes concluded that the resistance of the superconducting lead wire of at least 10 11 times lower than its resistance in the normal state. Subsequently conducting similar experiments, it was found that the current decay time exceeds many years, and this indicated that the superconductor resistivity less than 10 25 ohms · m. Comparing this with the resistivity of copper at room temperature, 1.55 x 10 -8 ohm-m - the difference is so large that one can safely assume that: the resistance of the superconductor is zero, it is difficult to call another monitor and modify the physical quantity which would be addressed in the same "ground zero" as the conductor resistance below the critical temperature.

Recall known from school physics course Joule - Lenz: the current I am flowing through the conductor with a resistance R it generates heat. At this consumed power $P = I2R$. As little resistance to metals, but it often limits the technical possibilities of different devices. Heated wires, cables, machines, apparatus, therefore, millions of kilowatts of electricity thrown to the wind. Heating limits the throughput power, the power of electric cars. Thus, in particular, is the case with electromagnets. Obtaining strong magnetic fields requires a large current, which leads to the release of an enormous amount of heat in the windings of the solenoid. But the superconducting circuit remains cold, and the current will circulate without damping - impedance equal to zero, no power losses.

Since the electric resistance is zero, the exciting current of a superconducting ring will exist indefinitely. The electric current, in this case, resembles the current produced by the electron orbit in the Bohr atom: it is like a very large Bohr orbit. Persistent current and the magnetic field generated by it cannot have an arbitrary value; they are quantized so that the magnetic flux penetrating the ring takes values that are multiples of the elementary flux quantum F on = h / (2e) = 2.07 10 15 Wb (h - Planck's constant).

Unlike electrons in atoms and other micro-particles, whose behavior is described by quantum theory, superconductivity - macroscopic quantum phenomenon. Indeed, the length of the superconducting wire through which flows persistent current can reach many meters or even kilometers.

Thus, carriers it describes a single wave function. This is not the only macroscopic quantum phenomena. Another example is superfluid liquid helium or substance neutron stars.

ELECTRICAL RESISTANCE
SUPERCONDUCTORS

No experimental methods are fundamentally impossible to prove that any value, in particular, the electrical resistance is zero. It can only show that it is less than a certain value determined by the measurement accuracy.

The most accurate method of measuring low impedances consists in measuring the current decay time induced in the closed circuit of the test material. A decrease in current time energy LI 2 /2 (L - inductance loop rate) consumed for Joule heat: here integrating (I 0 - current value at t = 0, R - resistance of the circuit).

The current decays exponentially with time, and the electrical resistance determines the attenuation rate (at a given L).

For small R formula can be written as

Here dI- current change during Δt. Experiments conducted using a thin- walled superconducting cylinder with extremely small values of L showed that the superconducting current is constant (with accuracy) within a few years. It followed that the resistivity in the superconducting state is less than 4 · 10 25-ohm m or more than 10 17 times less than the resistance of copper at room temperature. Since the possible decay time comparable to the time the existence of humanIty, we can assume that R DC in the superconducting state is zero.

Thus, the superconducting current - it is only like a real-life example of perpetual motion on the macroscopic scale!

When R = 0, the potential difference V = IR on any segment of the superconductor and hence the electric field E inside the superconductor is zero. The electrons which create a current in the superconductor, moving at a constant speed without being scattered by the thermal vibrations of the crystal lattice atoms and its irregularities. Note that if E is not equal to zero, electrons carrying the superconducting current to accelerated without limit, and the current could reach an infinitely large value, which is physically impossible.

The situation changes if the superconductor is applied to the variable potential difference creates a variable superconducting current. During each period, the current changes direction. Consequently, the superconductor must exist in an electric field, which periodically slows down the superconducting electrons and accelerates them in the opposite direction. Since it consumes energy from an external source, the electrical resistance of alternating current in a superconducting state is zero. However, because the electron mass is very small, power loss at frequencies less than 10 10 - 10 11 Hz is negligible.

TUNNEL EFFECTS

In 1962, an article appeared before anyone unknown author B. Josephson, who theoretically predicted the existence of two extraordinary effects: steady and unsteady. Josephson theoretically studied the tunneling of Cooper pairs from one superconductor to another through any barrier. Before proceeding to the first Josephson effect, briefly the tunneling of electrons between the two metal parts separated by a thin dielectric layer.

The tunnel effect is known in physics for a long time, this is a typical problem of quantum mechanics. Particle (for example, an electron in the metal) approaches the barrier (e.g., a dielectric layer), to overcome which she classical ideas cannot, as its kinetic energy is insufficient. However, the area of the barrier with his kinetic energy could well exist. On the contrary, according to quantum mechanics, the barrier passage possible. The particle may have a chance, as it were, to pass through the tunnel through a classically forbidden region where its potential energy would be more like a full, i.e., the classical kinetic energy as it is negative. Quantum mechanics for the microparticles (electron) holds uncertainty relation $\Delta h \Delta r > h$ (x - coordinate of the particle, p - its pulse). When a small uncertainty of its coordinates in a dielectric $\Delta h = d$ (d - thickness of the dielectric layer) leads to large uncertainty, its pulse $Dp \geq h / \Delta x$, and consequently, the kinetic energy $p \, 2 / (2m)$ (m - a mass of particles), the energy conservation law is not violated. Experience shows that indeed between two metal electrodes separated by a thin insulating layer (tunnel barrier), electric current can flow the greater, the thinner the dielectric layer.

Josephson Effect

Physical objects in which the Josephson effect takes place, now called Josephson junctions or Josephson junctions or Josephson elements. To imagine the role played by Josephson elements in superconducting electronics, it is possible to draw a parallel between them and the semiconductor p-n-junctions (diodes, transistors) - element base conventional semiconductor electronics. Josephson junctions are some weak electrical communication between two superconductors. This linking can be made in several ways. The most commonly used types in practice weak link - is 1) tunnel junctions, in which the bond between the two film superconductor is carried out through a very thin (tens of angstroms) insulating layer between them - SIS-structure; 2) "sandwiches" - two film superconductor interacting through a thin (hundreds Angstroms) layer of a normal metal therebetween - SNS- structure; 3) the structure of the bridge type, which is a narrow superconducting bridge (bridge) of limited length between two massive superconducting electrodes.

Bearers overcurrent superconductors at $T = 0$ K are all conduction electrons n (0) (electron density). When the temperature rises appear elementary excitation (normal electrons) so that the concentration of n s of the superconducting electrons at a temperature T $n s (T) = f (0) -n n (T)$,where n n (T) - the concentration of electrons at a normal temperature T. In the Bardeen-Cooper-Schrieffer (BCS) for $T \rightarrow T$ c (critical temperature) $p s (T) \approx \Delta 2 (T)$where 2 D (T) - the width of the energy gap in the spectrum of the superconductor. All superconducting electrons form pairs associated state, known as Cooper pairs of electrons.

Cooper pair combines two electrons with opposite spins and pulses and therefore has a zero net spin. Unlike normal electrons having a spin of 1/2, and therefore obey Fermi-Dirac statistics, Cooper pairs obey Bose- Einstein statistics and condensed at one, the lower energy level. A characteristic feature of Cooper pairs is their relatively large size (about 1 micron) is much greater than the average distance between pairs (of the order of the interatomic distances). Such a strong spatial overlap pairs mean that all of the (condensation) of the Cooper pairs is coherent, which is described in quantum mechanics, the wave function of a single W =. DELTA.E ix. Here A - amplitude of the wave function, the square of which characterizes the concentration of Cooper pairs, h - the phase of the wave function, i - imaginary unit, P - -1. In the case of normal electrons, which are fermions, the Pauli exclusion principle, the electron energy is never exactly equal to each other. Therefore, from the Schrodinger equation for these particles, it follows that the phase velocity dq / dt of the wave functions of electrons normal differ; thus, phase h are uniformly distributed in the trigonometric circle, and the summation over all particles explicit dependence on h disappears.

The presence of a weak electric connection between the superconducting electrodes due to poor overlap of the wave functions of the Cooper pairs of electrodes, whereby such contact is also superconducting, but the density of the critical current value is much (by several orders of magnitude) smaller than the critical current density of the electrodes j c $\approx$ 10 8 A / cm 2. For tunneling structures and structures of the sandwich-type critical current density of Josephson, junctions-ing is typically in diapazonej jc from 10 1 to 10 4 A / cm 2, and their area SB within modern technology can be made from a few hundred to a few square microns. Therefore, the critical current of the Josephson element $I_c = j_{jc}. S$ may be from a few milliamperes to a few microamperes.

THE UNIVERSE

There are many aspects of computation in the universe; there are many phenomena in the universe that scientists have been able to transform them into mathematical formulas. For example, to demonstrate that there is a mathematical relationship between temperature and volume, the higher the temperature of the ball, the greater its size, and because of the density, we demonstrate who will float on the surface of the water, Archimedes used water and objects to discover this. He managed to convert all of this into a mathematical equation between mass and volume. Albert Einstein was also able to link mass, energy, and speed with a mathematical formula.

Many examples indicate this. Several theories have emerged that discuss the possibility of computerization of the universe, meaning that the universe is a huge quantum computer, so what is computing? This has always been a pivotal question in the field of computer science.

At the beginning of 1930, computing meant the function of people who used to operate supercomputers. And at the end of 1940, computing was defined as: a set of steps implemented by automatic computers to produce knowledge outputs, and this standard definition remained for fifty years after that, but now it faces many challenges ,as people from many fields began to accept the idea that computational reasoning is a way to understand science and engineering.

The internet is full of services that do computing without stopping. Researchers in the fields of physics and biology claim to discover computing processes from nature unrelated to computers; computing is now in all sciences. Computer scientists design build and program computers. But again, we go back to our question, what can be considered a computer? What is the rock missing (like a physical system) and owned by a computer?

According to Seth Lloyd, author of Programming the Universe, all the interactions that occur between molecules in the universe transmit not only energy, but also information; in other words, molecules not only collide, but perform mathematical operations, the entire universe computes, each an atom, an electron, and an elementary particle that keeps a lot of information, and every time two molecules collide, these two bits are processed.

By delving into the computing power of the universe, we can build quantum computers that store and process information at the level of atoms and electrons. This computing power forms the basis of complex systems and provides a deeper understanding of the origin and future of life.

Computing is taught in two ways: theoretically and materially; Mathematics is concerned with studying the theoretical side of computing by providing mathematical definitions of computational matters such as algorithms and providing theories about their properties.

One of the most important questions was: is first-order logic decidable? That is, is there an algorithm that can decide whether a particular logical phrase (first-order) is a theory? Turing and Church proved that the answer is negative; there is no algorithm with this description.

To demonstrate this, they provided an accurate description of the unknown concept of a computing function and writing it in an algorithm.

Turing did this through what is known as a Turing machine, which is a virtual machine that processes separate symbols written on a tape compatible with a limited number of commands. The study of computing functions is made possible by the work of Turing and others.

According to the Turing-Church hypothesis, any process that is intuitively computing is computing with the Turing machine; this can be formulated as follows: "Any function that is seen as naturally computing is a Turing-computable function."

Intuitively computable, which means that it is computable by following an algorithm or an effective procedure. The practical approach includes an expired list of clear commands for the production of new symbolic structures based on ancient symbolic forms. Some scholars study that the physical universe is a computational basis, the universe itself is a computational system, and everything inside it is a computational system as well, where it is viewed according to two different perspectives, the first of which is an automatic operator that represents the traditional computational model and the quantitative, non-quantitative.

The idea that the universe could be a giant digital computer existed decades ago. In 1960, Edward Fredkin, then Professor of the MIT Institute Konrad Zuse, who built the first electronic digital computer in Germany at the beginning of 1940, proposed the idea that the universe is an integrated digital computer (and recently, the idea was supported by computer scientist Stephen Wolfram.

According to these physicists, the universe is a giant cellular robotic operator. The robot is a network of cells, with each section taking a state from within a finite set of conditions and updating its form with separate steps according to the adjacent countries. For the universe to be a cellular robotic, all physical amounts must be separate. Besides, time and space must be separated. Although cellular robotics are capable of describing many basic physical phenomena, the quantum features of the universe are challenging to simulate using a conventional model such as cellular robotics.

The universe is quantum, and standard computers cannot simulate quantum systems; why? Because quantum mechanics is more exotic and counterintuitive compared to regular computers as for humans to affect a tiny part of the universe, to transfer a few hundred atoms to one part of a second, a regular computer needs more memory space than the number of atoms in the entire universe, and to a longer time than the present age of the universe to end the simulation.

This is what led to the development of quantum computing models. Instead of relying on numbers - often numbers or bits - quantum computing relies on qubits. The difference between qubits and bits is that while bits can take one value from one of two values: 0 or 1, qubits can take a set of values that represent the superposition of states 0 and 1.

According to this principle, the universe is not a classic computer but rather a quantum computer; that is, it is a computer that does not process numbers but qubits. The quantum version of the universe is less radical than the traditional version, since the conventional version eliminates continuity from the universe, by claiming that deleting it allows classic computers to provide a literal description of the universe rather than an estimated one.

The universe is a physical system that is subject to computation; therefore, it can be simulated effectively using a quantum computer - the size of the universe itself - and because the universe supports quantum computing. It can be manufactured using a quantum computer. It possesses a computing power that is no less than a quantum computer of the universe's size. We have seen how physics laws can be used to perform quantum computing effectively; let us discover how a quantum computer can simulate physical laws.

Quantum simulation is the process by which a quantum computer simulates a quantum system. Because of the strange quantum properties, classic computers affect quantum systems less effectively. Still, because the quantum computer is itself a quantum system capable of highlighting all quantum properties, it can effectively simulate other quantum systems. Each part of the quantum system that you want to emulate is stored within a group that is stored within a group of the qubits are inside the quantum computer. The different interactions between these parts are transformed into logical quantum computational operations. The resulting simulations are so accurate that it is difficult to differentiate them from the simulated system.

No description of the universe was found as a computer before the twentieth century. The ancient Greeks atom indeed counted the universe as a form of interaction of small parts, but they did not explain whether these parts were information processing units.

Laplace invented a virtual object that calculates the future of the entire universe but considered it a separate entity from the universe and not the universe itself.

At the same time, Charles Babbage was not eager to use his device as a model for physical phenomena, unlike Alan Turing. The latter was interested in the origin of patterns and researched the topic.

The first explicit description of the universe as a vast computer was in 1956 in the science fiction novel "The Last Question" by Isaac Asimov. In this story, humans invent analog computers to help them explore their galaxy first and then other galaxies. The link between computing and physics began in the early 1960s by Rolf Landauer at IBM. The idea that the universe can be regarded as a computer was proposed by Fredkin and independently by Konrad Zuse. Frieden and Zeus suggested that the universe could be a type of classical computer called a cellular robot that contained rows of bits that interact with adjacent bits. Stephen Wolfram recently developed and simplified this proposal. The idea of using cellular robotics as a basis for the theory of the universe seems interesting. Still, its problem is that classic computers are not able to reproduce quantum phenomena, such as quantum entanglement.

Another reason, as mentioned earlier, is that simulating a small portion of the universe on a classic computer requires a volume equal to the size of the universe.

Therefore, it is impossible to regard the universe as a classic computer as a cellular robot. In his research paper: Ultimate Physical Limits to Computation, Seth Lloyd demonstrated how the computing power of any physical system could be calculated by knowing the amount of energy present in the design and the size of the system, for example: Use these limits to calculate the maximum computing power of one

kilogram of material stored in a liter volume one from space, the standard laptop weighs approximately one kilogram, and takes the size of one liter of hole so that we will call the one-kilogram computer and one liter of super-laptop.
A super laptop is a 1-kg computer and 1 liter (regular laptop size), where every elementary particle was placed inside it for computing.
A super laptop can perform 10 million logical calculations per second at ten thousand billion bits. What might be the power of a super laptop?
The first significant impediment to excellent computational performance is energy. The amount of energy determines the range of speed, for example: Let's take a single-bit electron, which moves here and there, whenever he has much energy whenever he moves quickly here and there and can change his bit-flip state quickly, the speed at which qubits change his condition is subject to a theory known as Margolus-Levitin. The theory says that the maximum speed at which a specific physical system (for example, an electron) changes according to its energy.

Examples And Applications

The theory, like classical mechanics, deals with the motion of particles in space and time. The difference only lies in that classical mechanics describes the deterministic continuous motion of particles, while quantum mechanics describes the random discontinuous motion of particles.

Although the new formulation of quantum mechanics makes the theory as comprehensible as classical mechanics, the quantum phenomena are still strange for us who live in a classical world. In this part, we will give several examples to illustrate that the weirdness of the quantum world, which is lacking in the everyday world, all originates from the new motion of particles and its new laws, e.g., the discontinuity and randomness of motion, the superposition principle, the collapse of the wave function, etc.

These examples may help people understand quantum mechanics more deeply.

1
2
3a
3b
Schrödinger's cat

SCHRÖDINGER'S CAT

In 1935, inspired by the famous paper written by Einstein, Podolsky, and Rosen, Schrödinger proposed the more famous Gedanken experiment, which is called Schrödinger's cat paradox. The experiment was described by Schrödinger as follows:

A feline is written up in a steel chamber, alongside the accompanying insidious gadget (which must be made sure about against direct obstruction by the feline); in a Geiger counter, there is a smidgen of a radioactive substance, so little that maybe in one hour one of the iotas rots, yet additionally, with equivalent likelihood, maybe none; on the off chance that it occurs, the counter cylinder releases and through a transfer delivers a mallet which breaks a little carafe of hydrocyanic corrosive. On the off chance that one has left this whole framework to itself for 60 minutes, one would state that the feline, despite everything lives assuming then, no iota has rotted. The main nuclear rot would have harmed it. The ψ-capacity of the whole framework would communicate this by having in it the living, and the dead feline (pardon the articulation) blended or spread out in equivalent parts.

So, when a rotting molecule cooperates with a feline utilizing a lot of gadgets, including an indicator, a sled, and a little carafe of hydrocyanic corrosive, and so on., the entire framework will be in an entrapped superposition containing two branches as per the direct Schrödinger condition. In one branch, the particle rots and toxins the feline, while in the other branch, the molecule doesn't rot, and the feline, despite everything, lives. As an outcome, the feline will be in a superposition of dead state and living state. Let's first see what Schrödinger's cat looks like according to the picture of the random discontinuous motion of particles. The following figure depicts Schrödinger's cat at six neighboring instants. At each instant, the cat is either alive or dead in a purely random way.

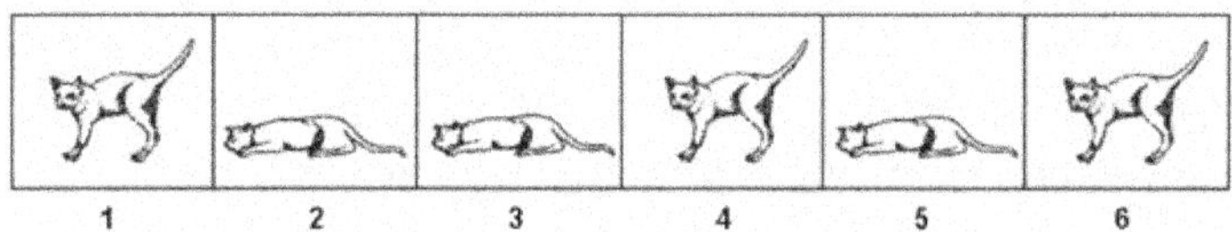

Figure1.3 Instantaneous pictures of Schrödinger's cat
However, during a time interval, the random discontinuous motion of the cat will form a cat cloud, which is quite like an electron cloud. In one branch, the cat is alive, and in the other, the cat is dead.
The problem is how to reconcile the bizarre state of Schrödinger's cat with our macroscopic experience. This is essentially the measurement problem. The new formulation of quantum mechanics provides a clear solution to this problem, though some details are still missing. In Schrödinger's cat experiment, the atom is indeed in a superposition of the undecayed state and the degraded state. The evolution of the state is also governed by the linear Schrödinger equation very precisely, as the effect of the stochastic evolution is extremely small, e.g., the collapse time of this superposition state may be longer than the age of our universe. However, when the atom interacts with the Geiger counter or the cat and its superposition state is entangled with the state, the effect of the stochastic evolution becomes significant. Thus the whole superposition almost immediately collapses into one of its branches. As a result, the Geiger counter and the

cat will always be in a solid-state; the Geiger counter either registers a particle or registers no particle, and the cat is either living or dead.

QUANTUM ENTANGLEMENT AND NONLOCALITY

In this part, we will present a clear physical picture of quantum entanglement and nonlocality, which are widely regarded as the most puzzling phenomena of the quantum world.

Let's first recall the single-particle picture. For the arbitrary intermittent movement of a molecule, the molecule inclines to be in any conceivable situation at a given moment. The likelihood thickness of the molecule showing up in each position x at a given moment t is dictated by the modulus square of its wave work, specifically $\rho(x, t) = |\psi(x, t)|2$. The physical image of the movement of the molecule is as per the following. At the moment, the molecule haphazardly remains in a position. At the instant, it will still stay there or randomly appear in another position, which is probably not in the neighborhood of the past position. In this way, during a time interval much larger than the duration of one discrete instant, the particle will move discontinuously throughout the whole space with position density $\rho(x, t)$. Since the distance between the locations occupied by the particle at two neighboring instants may be very large, the jumping process is nonlocal. In other words, two staying events of the particle (t1, x1) and (t2, x2) may readily satisfy the space-like separation condition $|x2- x1| > c|t2- t1|$.

Let's turn to the motion of two entangled particles. For the random discontinuous motion of two particles in an entangled state, the two particles have a joint propensity to be in any two possible positions. The probability density of the two particles appearing in each position pair x1 and x2 at a given instant t is determined by the modulus square of their wave function at the instant, namely $\rho(x1, x2, t) = |\psi(x1, x2, t)|2$.

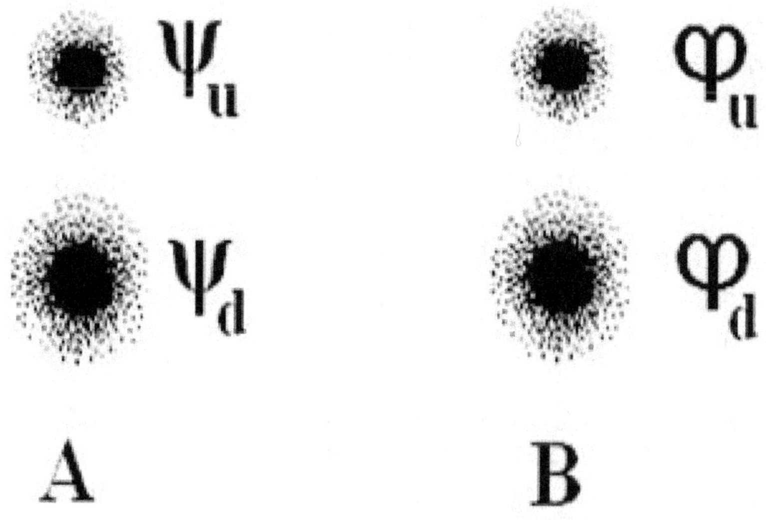

Figure 1.3 Two particles in an entangled state

Assume two particles are in an ensnared state $\psi_u\varphi_u+\psi_d\varphi_d$, where ψ_u and ψ_d are two spatially isolated conditions of molecule 1, φ_u and φ_d are two spatially isolated conditions of molecule 2, and molecule one and molecule two are additionally in isolated areas An and B. The physical image of this trapped state is as per the following. Particles 1 and 2 are haphazardly in the state $\psi_u\varphi_u$ or $\psi_d\varphi_d$ at the moment, and afterward, they will even now remain in this state or hop to the next state at the moment.

During a brief timeframe stretch, the two particles will spasmodically move all through the states ψuφu and ψdφd with a similar likelihood 1/2.

Along these lines, the two particles structure an indistinguishable entire, and they go through synchronous hopping. At an emotional moment, if molecule 1 is in the state ψu or ψd, at that point molecule two must be in the stateφuorφd, and the other way around. Also, when molecule one hops from ψu to ψd or from ψd to ψu, molecule two should all the while hop fromφutoφdor fromφdtoφu, and the other way around. Note that this kind of random synchronicity between the motion of particle one and the motion of particle 2 is irrelevant to the distance between them, and it can only be explained by the existence of the joint propensity of the two particles as a whole. This gives us a clear physical picture of quantum entanglement.

Thirdly, let's analyze the nonlocal characteristic of the collapse of the above-entangled superposition during a measurement. The property of quantum nonlocality was first brought to the attention of the physics community by Einstein, Podolsky, and Rosen in 1935. John Bell made a big stride forward in the study of quantum nonlocality in the 1960s. The famous Bell theorem shows the contradiction between locality and quantum mechanics (especially its collapse postulate). Many experiments have been conducted to test Bell's theorem up to now. Although the results confirm the predictions of quantum mechanics and reveal the actual existence of quantum nonlocality, its nature is still a mystery.

As Bell once said, "The scientific attitude is that correlations cry out for explanation." Now the random discontinuous motion of particles and its laws may provide a true explanation of nonlocal quantum correlations.

Suppose we make a position measurement on particle 1 in region A. Let the initial state of the measuring device be φ0. When the measuring device interacts with particle 1 in a local region, the state of the measuring device will be entangled with the entangled state of particles 1 and 2, and the state of the combined system will become ψuφuφu+ψdφdφd. Since the stochastic evolution of the state will take effect during the measurement process, the state of the combined system will soon collapse to one of the branches ψuφuφu or ψdφdφd with the same probability 1/2. As a result, the combined system as an entangled whole is disassembled into three independent parts: the measuring device, particle 1, and particle 2. At the same time, the original two-particle whole no longer exists either, and its state collapses to ψuφu or ψdφd, i.e., the state of particle one collapses to ψu or ψd, and the state of particle 2 collapses toφuorφd. It can be seen that the local measurement of the measuring device brings a nonlocal influence on the two-particle system and especially on particle 2.

UNIFICATION OF TWO WORLDS

We live in a classical world. The objects around us appear to move continuously. In the quantum world, however, every particle moves in a purely random and discontinuous way. If the motion of all objects is essentially discontinuous and random, then why does the motion of macroscopic objects appear continuous? In this part, we will briefly explain how the transition from quantum to classical happens and why the random discontinuous motion of particles may provide a uniform picture for both the microscopic and macroscopic worlds.

We will first state the laws of random discontinuous motion more explicitly. Although the complete laws of motion are still unknown, we can formulate its general form. According to our past analysis, the (nonrelativistic) evolution of the wave function will be governed by a revised Schrödinger equation that contains two kinds of evolution terms. The first is the deterministic linear Schrödinger evolution term, and the second is the stochastic nonlinear evolution term that results in the dynamical collapse of the wave function. The equation can be formally written in a discrete form:

$$\psi(x, t+T_p) - \psi(x, t) = \frac{1}{i\hbar} H \psi(x, t) T_p + S \psi(x, t).$$

where Tp is the Planck time, the duration of a discrete instant, H is the Hamiltonian of the system for the linear Schrödinger evolution, and S is the stochastic evolution operator for the stochastic nonlinear evolution. Note that all quantities in the evolution equation should be defined in discrete space-time.

In the complete evolution equation, the linear Schrödinger evolution term will lead to the spreading of the wave function. In contrast, the

nonlinear stochastic evolution term will lead to the collapse or localization of the wave function. If the energy of the system is very small, then the evolution will be dominated by the spreading process. This is just what happens in the microscopic world; a particle can pass through two slits sat the same time in the double-slit experiment.

THE QUANTUM WORLD OF SPINNING PARTICLES

ANGULAR MOMENTUM IN CLASSICAL MECHANICS

Before focusing on the mysteries and paradoxes of the quantum world, we must learn a couple of elementary concepts – such as angular momentum and spin – that we will find throughout the rest of this reading. So, what is angular momentum? In CP, it is not very hard to understand. It is the amount of rotational momentum, the quantity of motion of rotating bodies. Therefore, let us first look at the notion of the angular momentum of a single particle rotating around a center, as shown diagrammatically, with vectors in Fig. 2.1.

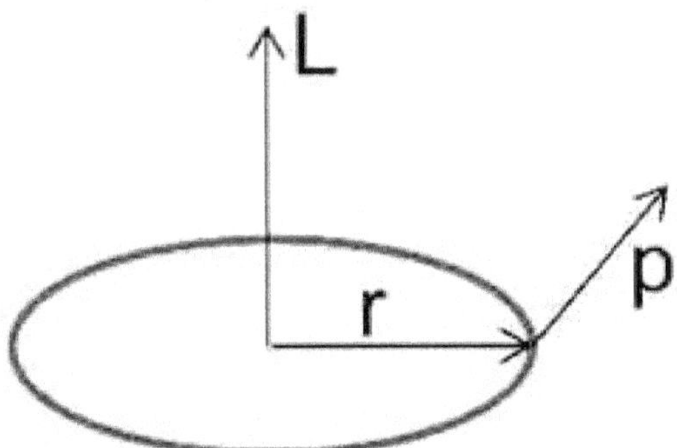

Fig. 2.1 The angular momentum of a particle.

Recall how the linear momentum of a particle that moves unperturbed along a straight path is proportional to the product of its mass times velocity (at least in non-relativistic CM): = · . Similarly, the

general expression of the angular momentum of a particle with rotational velocity about an origin is defined as:
= × = × · , Eq. 1

Where the position vector of the particle relative to the rotational origin, and the cross denotes the vector or cross-product (a sort of multiplication for vectors).

Please note that both the linear and the angular momentum are vector quantities, which means they have a magnitude and direction. Letters representing vectors are written with an arrow or in bold.
(Here, we use the latter.) The angular momentum vector L is perpendicular to the r x p plane. Its magnitude (length) is a measure of the rotational speed. In the event of a body made of n particles rotating around a center external to itself (that is, orbiting around a center of origin), the 'orbital angular momentum, can be obtained by determining the sum of the linear momenta of each of its constituent particles – the sum over all the particles of masses times the respective rotational velocity for each, as:
= Σ ×, again with as the i-the particle's distance from that axis and n as the number of particles. However, an extended body can also rotate around itself.

A typical example is that of the Earth, which not only orbits around the Sun but (in case you didn't notice) spins once around its polar axis approximately every 24 hours.

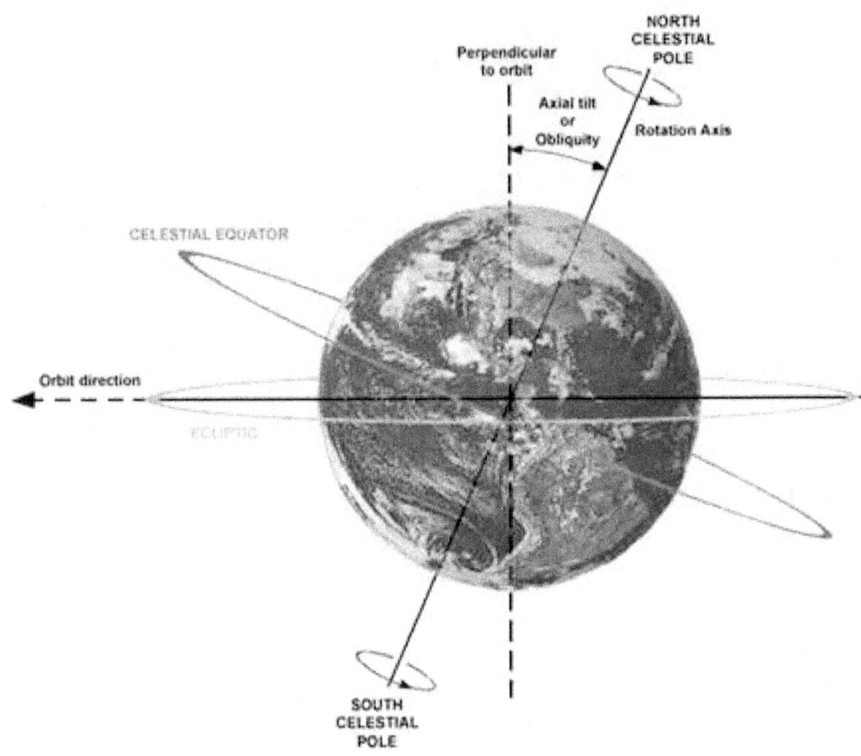

Fig. 2.2 The Earth's rotation.

In this case, for bodies whose interiors have complex mass distribution, like the Earth (the density distribution inside the earth can be quite complicated), one must calculate the so-called 'moment of inertia' of a body.

Inertia is a physical quantity that measures the resistance a material body exerts upon the change of motion. In the case of a point-like particle, it is simply its mass. In the case of a body with an extension and a more or less irregular complex geometry and internal matter distribution, it is a scalar quantity, a number I, which can be calculated using mathematical procedures.

In general, the 'spinning angular momentum' is a vector given by:
= ·Eq. 2
Where is the mathematical expression of the 'angular velocity' which is the measure of how fast it spins around itself – that is, how quickly it completes a 360° rotation per unit time, which, in the case of a spherical body, is just =, the rotational speed of the body divided by its radius.

Therefore, we must distinguish between the angular momentum of a particle or a body moving around a center (e.g., a planet around the Sun) and the angular momentum of a body around its central axis (e.g., the case mentioned above of the angular momentum of the Earth spinning around its polar axis). These are two closely related but slightly different quantities. One is an orbital angular momentum, while the other is a spinning angular momentum, also known simply as 'spin.' The total angular momentum is the sum of the two:
= +.

An important universal law to keep in mind is that of the 'conservation of angular momentum,' which is a direct consequence of the conservation of energy (i.e., energy can never be created or destroyed, but only transformed).

 In Eq. 1 the radius or the speed are allowed to change, but the overall momentum L must remain constant. If a particle gets closer to the rotation center (r decreases), the angular velocity must increase to maintain constant. This also holds for any extended body. A common example illustrating this principle is that of the ice-skater. When ice- skaters bring their arms closer to their bodies, the angular velocity increases, and vice versa. This act, for example of the ice-skater, decreases the inertia I; according to the law of conservation of angular momentum, an increase of must follow to maintain constant (see Eq. 2), and vice versa.

Fig. 2.3 Conservation of angular momentum
Now that we have been introduced to the main concept and principle of the angular momentum of CM let us apply this to QM.

Spin, The Stern-Gerlach Experiment, And The Commutation Relations

In QM, one is typically more concerned with single particles or many- particle systems than with extended bodies. So, what about elementary particles, like electrons? It is not difficult to extend the notion of the orbital angular momentum of a material particle having a mass like, for example, the electron in the case we conceived of it in Bohr's atomic model, as a particle flying around the atom's nucleus.

This conception is dubious enough because we should have in mind the atomic orbital model, not Bohr's atomic model. In any event, so far, this picture of reality still works: The orbital angular momentum of the electron can be defined as the product of its mass times the velocity at which it orbits the nucleus times the orbital distance from it (Eq. 1 - this was also Bohr's mathematical approach which led to his atomic model).
However, this is where the analogies to the classical world end.

In fact, what about the spin – that is, what in QM is called the 'intrinsic angular momentum' – of a particle rotating around an axis that goes through its center, as in the case of the earth? The problem is, this analogy breaks down because we think of so-called 'elementary' particles as a mathematical point-like object. Thus, it becomes somewhat unclear what the spin of a point structure should mean at all. What is a point that spins around itself ?
Intuitively and even mathematically, it doesn't make much sense.

So, either electron is not point-like, but no experiment
has so far revealed any internal structure (at least not
as of 2020), or we must consider the notion of the
spin of particles in QM as something that must not be
viewed as we do intuitively at the macroscopic scale.
For this reason, in QM, one more loosely speaks of
particles 'carrying' an intrinsic angular momentum, or
spin, without suggesting the image of a spinning
sphere literally, as in classical theory.

And yet, we know that even elementary particles such
as electrons, protons, and neutrons have a tiny but
measurable spin. Can the spin of a particle as tiny as
an electron be measured? The answer is positive and
even surprising. It is possible to measure and
associate spin with elementary particles. Interesting is
that, as we saw energy being quantized in QM, the
spin of particles also turns out to be quantized. Nature
allows only discrete– not continuous – quantities of
intrinsic angular momentum! Spin in QM is a
quantized dynamical property of all particles.

How physicists reached this conclusion will be
elucidated with the Stern- Gerlach experiment. It is
easier to understand how this works if we first
introduce some basics and anticipate some of the
results.

For elementary particles, the spin has only two
possible values. If we take as a convention the vertical
axis (see Fig. 2.4) being the upward z- axis, we can
show experimentally that, say an electron has only
two possible spins along that axis: a fixed specific
amount of spin-up or spin-down states. We will never,
ever observe the electron having other amounts of

spin and being directed towards another direction in between.
Or, if you would like to stick to the classical intuitive understanding, elementary particles rotate clockwise or counter- clockwise around an axis, always with the same angular frequency.

What must be added to this picture is the fact that we should not forget that several particles, like electrons and protons, are electrically charged particles. (Neutrons are not.) They all possess an equal tiny charge, which is also taken as the elementary electric charge, labeled 'e,' which amounts to

$$= 1.672 \ 10$$

C, where [C] stands for 'Coulomb' and is the standard international unit for the electric charge. This is an extremely small charge.

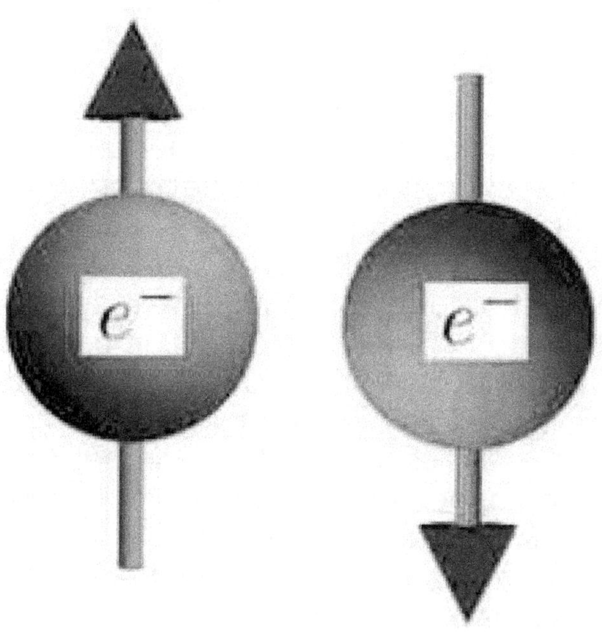

Fig. 2.4 A spin-up and spin-down electron.
For example, a typical household electric device (say,
a small lamp) that works with an electric current of
1A ([A] stands for 'Ampere,' the unit of the intensity
of an electric current), has something of the order of
6x10 (six quintillions, that is, six billions of billions)
electrons per second flowing through its circuitry.

However, despite its small value, the electric charge of
an electron can be measured using relatively simple
considerations and experimental arrangements.

Electrically charged particles produce an electric field
in their surrounding space, which interacts with other
charged particles or with magnetic fields. They are
also themselves the source of magnetic fields.

Every moving electrically charged particle always
produces a corresponding magnetic field. A cable
through which an electric current
–   that is, electrons
–   flows will also manifest a magnetic field.
This is always true and is a fundamental law of
physics: Wherever electric particles travel throughout
space or through a conductor, they will always
produce a magnetic field.
(The construction of electromagnets relies on this
principle.) A loop of electric current, a bar magnet, an
electron, a molecule, and a planet all have magnetic
moments, as they all contain, in one form or another,
circulating electric charges. This is something we
imagine also happening with the Earth's magnetic
field. The Earth's interior must contain a huge amount
of magma, of hot currents of a magma fluid, which is
electrically charged. Its movement around the Earth's
axis causes the Earth's magnetic field to build up. The

opposite is true, as well. Wherever a magnetic field changes in time, an electric field will appear. This is what we already hinted at when discussing the nature of light as an oscillating EM field, and it is at the foundation of Maxwell's equations.

Therefore, an electrically charged spinning object is also expected to display some magnetic field, as it is a charge in rotational movement. The charge flow around itself induces a particle's magnetic field due to its spin. This makes every elementary particle, like electrons and protons, also tiny magnets. Therefore, not only do they possess an electric charge, but because they possess a spin, they must produce a small magnetic field as well.

And because only two fixed spin states are possible, the result is a 'magnetic moment,' with the magnetic field lines directed in one or the other direction according to the particle's spin orientation, as shown in Fig. 2.5.

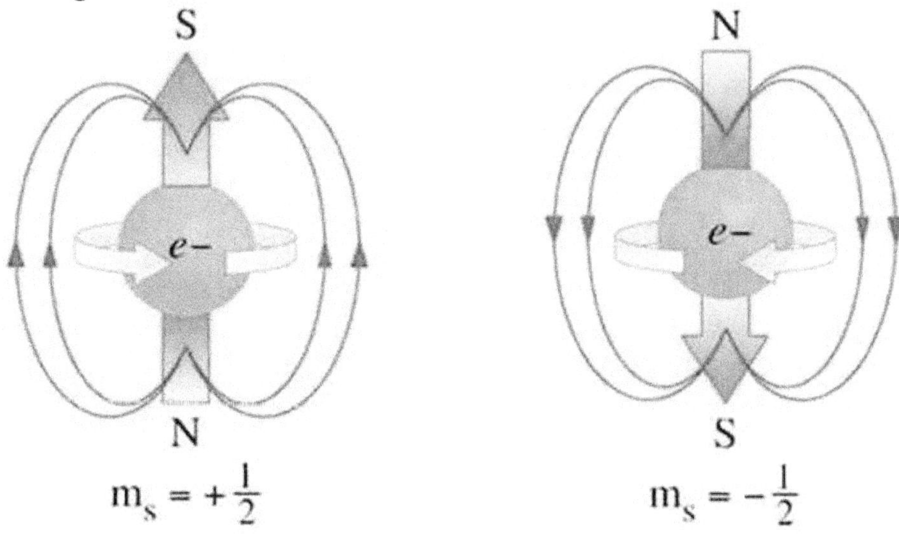

Fig. 2.5 Spin orientation and magnetic field lines of an electron.

MODERN COMPLICATION

Physicists can have it rough when confronted with mysterious problems, not because they have any more issues than they would have if the problems weren't mysterious, but because popular culture tended to convolute an understanding of the mystery. It is no surprise that the media and movies can twist science when science does not specify precisely what it is talking about. It has been speculated that there is a different type of matter in our universe that accounts for certain mysterious observations. Galaxies have been heavier than they should be, stars move faster than they should move, and particles collide unexpectedly. A string of explanations has been provided by research scientists in the area of astronomy. In general, these explanations incorporate the existence of dark matter.

The reason why dark matter is "dark" is that it is invisible to the human eye and every telescope technology built thus far. What that means for research purposes is that dark matter cannot be as easily experimented with, as can particles that fall under the definition of "matter." If you want to test Newton's laws of gravitation, you can take a banana, put it next to Venus, and measure the gravitational force on the banana as a function of whatever variable you want. You can look at the banana's orbital acceleration, for example, but you cannot find yourself a dark banana, put it next to Venus, and understand how it behaves in response to varying physical parameters.12

12 No, it does not matter how ripe the banana is.

 The universe consists of 27% dark matter, which poses a very interesting question: how is it that over one-fourth of the universe is comprised of something we can't even see? Moreover, how has there been so much experimental support for something we can't work with?

A Dark Matter Galaxy?

The answer is that you don't need to see something to "do science" on it. Two branches of physics, in particular, have proven very promising areas of experimentation: astrophysics and particle physics. Very recently, a team of researchers led by Peter van Dokkum at the W. M. Keck Observatory in Hawaii discovered an intriguing galaxy named Dragonfly 44. Dragonfly 44 has a very small collection of stars. Van Dokkum's objective was very simple: he knew the stars would be moving at some orbital speed. Based on previous literature, however, he also understood that certain stars in Dragonfly 44 moved far too quickly for what their actual speed should be, according to calculations. The strategy was to evaluate what the orbital speed was supposed to be and measure what the speeds of the stars were. Comparative analysis between these values might reveal interesting (or maybe not!) possibilities.

The expected orbital speed of the stars is not incredibly difficult to calculate. Two theories of gravitation can be used to do so: Newtonian's law of gravitation or Einstein's theory of general relativity. According to the former, the more matter there is in a system, the larger the magnitude of the force on each component. The larger the force, the greater the acceleration, and the greater the change in orbital speed. According to the latter, gravity is related to spacetime curvature, but there is a similar argument: more matter in the spacetime map leads to more curvature, meaning stronger gravitational effects. Both situations postulate that the more matter there is in a given area of space, the greater the change in orbital speed should be of the components (stars) that make up that system. Because there were relatively few stars in Dragonfly 44, it was assumed that the stars could not be moving very

fast. However, when they calculated experimental orbital speeds, they ended up with values that were significantly greater than what they should have been. It doesn't matter which team you're on, Newton or Einstein because you're ultimately going to come to the same conclusion: if these stars were moving faster than they should be moving, there has to be more mass in the system. Precisely, the mass that scientists before Van Dokkum could not detect.
It fit the dark matter description quite nicely. If Dragonfly 44 were composed of dark matter, it would have a mass that we can't measure. That extra mass might contribute to gravitational effects experienced by the stars, which is why they move with such fast orbital velocities. The next objective was to dig a bit deeper: if it could be calculated how much mass was necessary to force the stars to move at their speed, it could be known how much dark matter is present in this galaxy. It turned out. Dragonfly 44 is composed of dark matter by nearly 99%. Van Dokkum and his team found a dark matter galaxy.

There is a particle accelerator at the CERN laboratory that collides large hadrons or a specific class of particles. Despite nearly a decade of debate about a unique name for the collider, participating physicists settled on the Large Hadron Collider (LHC). In the LHC, particles can be crashed into each other to produce collisions. When two particles collide, two variables are important: the energy and momentum changes of the collision. Generally, in ideally isolated systems, both energy and momentum of collisions are conserved. That means if you send two bananas crashing into each other and watch the two bananas stick together after colliding, the energy of one banana added to the energy of the other banana will give you the total energy of the stuck-together bananas at the end. The same applies for momentum. Yet, that's not always the case. If baryons and fermions, members of other classes of particles, are sent moving towards each other, momentum and energy conservation is generally not observed. Energy and momentum tend to both be lost. However, if you force the baryons and fermions to collide ideally, meaning you deprive the system of everything but these two particles, you should get momentum conservation. That's exactly what has happened at the LHC: baryons and fermions have been put in isolated systems with the momentum before and after the collision both measured. What should have been no difference became a very obvious difference: momentum was lost.

It was, therefore, postulated that dark matter particles were produced in these collisions. Such dark matter particles carried away some momentum to produce the change that was measured. Because dark matter can slip from detection, necessary measurements to verify these postulations could not be made directly. When the momentum of these supposed dark matter particles was measured and compared to the momentum change of these particle collisions, the data were exceptionally telling.

Five Modern Application Of Quantum Physics

he concepts of Quantum Physics is difficult and strange. Understanding the activities of tiny particles and trying to define the forces that made them work led Albert Einstein and his colleagues into an argument about the subject. The issue with Quantum Physics is that it has a very strange concept that defies common sense notions on causality, locality, and realism. Realism makes us know that something exists- we could know the sun exists even without staring at the sun.

Causality explains that something happens because something caused it to. Flicking the switch of a light bulb and we see the light, that's causality. Due to the speed of light, when we strike a match, the light does not take a million light-years to come on; this is all dependent on the location. All of these principles are not followed in the quantum realm. It's a whole different world there.

One clear example of this is the quantum entanglement, and this states that particles on opposite sides of the universe can be entangled to exchange knowledge instantly. This was a concept Einstein couldn't accept. In the year 1964, a physicist by the name John Stewart Bell was able to prove that quantum physics was a complete and workable theory. He was able to define Bell's Theorem.

The Bell theory proposed some series of inequality, now referred to as the Bell inequality; this series represented how measurements of the spin of a particle A and Particle B would be distributed if they were not entangled.

 While the experiment was being carried out, and after it was carried out, it was discovered that Bell's inequality was violated. He has been able to show that quantum properties like entanglement are as real as staring at a tree.
And in this time, the various strange concepts of Quantum Physics have been applied to develop different systems with real-world applications.

Here Are 5 Of The Most Exciting Ones:

1.   Ultra-Precise Clocks

The need to have a reliable timepiece with precision is very necessary. Times are already synchronizing the technological world; the time helps to keep the stock markets as well as maintain the GPS systems. The standard clocks we know make use of frequent oscillations of physical artifacts such as pendulums or quartz crystals to create their 'ticks' and 'tocks.'
In today's world, the most accurate clocks make use of theories of Quantum mechanics to calculate time. They monitor the specific frequency of radiation, which is required to make electrons jump between energy levels.

The Quantum logic clock, which is located at the US National Institute of Standards and Technology (NIST) in Colorado, only gains or loses a second every 3.7 billion years. The NIST strontium clock, which was revealed not so long ago, will remain accurate for the period of five billion years, a time longer than the Earth's current age.

The importance of this super-accurate and sensitive clock cuts across different areas: telecommunications, GPS navigation, and surveying. The precision of these atomic clocks partially depends on the number of atoms being used.

When scientists try to cram about a hundred more atoms into an atomic clock, this will make the accuracy of the clock about ten times more.

2.   Uncrackable Codes

In the traditional form of cryptography, keys were used to make it work. The sender of the secret message uses a particular kind of key to encode the information, and the recipient of the information will be able to use another key to decode the information or message. However, the risk that this message can still be picked is one that cannot be taken for granted. To solve this problem, technologists have employed the use of a theoretically unbreakable distribution of the quantum key (QKD). In QKD, done polarized photons are used to send the main information. This limits the photon; this makes it vibrate in just a singular plane -it could be up and down or left to right.

The other party accepting this data would then be able to utilize captivated channels to unscramble the key and afterward utilize the picked calculation to scramble the message appropriately. Individual information is as yet sent over typical correspondence channels, yet one can just disentangle the message except if they have the specific quantum key. The quantum decides to give that "perusing" enraptured photons will consistently change their states, and any push to listen in will alarm the communicators about security penetrate. On this day, organizations, for example, Toshiba, BBN Technologies, and ID

Quantique, utilize the QKD to plan super safe systems. In 2007, Switzerland had the option to embrace an ID Quantique item to give a carefully designed democratic framework in their races. In 2004, Austria had the option to receive trapped QKD just because to make the bank move.

3.   **Super-Powerful Computers**

PCs, for the most part, encode their data as twofold digit string or bit string. Quantum PCs supercharge the handling power since they utilize quantum bits or qubits which exist in a superposition of states — until they are estimated, qubits can be both "1" and "0" at the same time. Even though the field is as yet being worked on, there remain hints of movement the correct way. D-Wave frameworks uncovered in 2011

their D-Wave One, which has a 128-qubit processor, and afterward, D- Wave Two turned out in the next year and bragged of a 512-qubit processor.
A report from the company says that these are the first commercially available quantum computers in the world. This has remained an issue because it is still unclear if D-Wave's qubits are entangled. Research published in May finds signs of entanglement, but only in a rather small subset of machine qubits. There is confusion as to whether the chips exhibit any real quantum speedup. There has been a recent collaboration between NASA and Google to develop a D-Wave Two Quantum Artificial Intelligence Lab. Scientists at the University of Bristol were able to connect one of their traditional Quantum chips to the Internet so that anyone with a web browser can learn Quantum coding.

4.   Improved Microscopes

Some research teams from Japan's Hokkaido University were able to develop the world's first entanglement-enhanced microscope, which made use of differential interference contrast microscopy technique. This particular kind of microscope burns two beams of photons to a material and tests the interference pattern changes depending on whether the beams touch a smooth or irregular surface. Using entangled photons considerably increases the amount of information that the microscope can collect, as measuring one entangled photon provides information about its partner. The Hokkaido was able to build an etched "Q," which was just 17 nanometers above the surface with unparalleled sharpness. Astronomy instruments such as interferometers can have their resolution increased using similar methods. The interferometers are used to search for extrasolar planets, to probe surrounding stars, and to search for space-time ripples or gravitational waves.

5.   Biological Compasses

The use of quantum mechanics isn't something used only by humans. It's being observed that birds such as the European Robin use the weird behavior to keep track of their migration. They carry this process out through a light-sensitive protein called cryptochrome, and this may have entangled electrons. As soon as the photons move into the eye, they reach the cryptochrome molecules. Enough energy is provided to break them apart, and this forms two reactive molecules or radicals with the unpaired but still entangled electrons.
The duration the cryptochrome would last is largely dependent on the magnetic field around the bird.

The birds have a very sensitive retina, which can easily detect the presence of entangled radicals; this enables the animals to notice a molecular-based magnetic map effectively.

Although, when the entanglement becomes poor, the experiment has shown that the bird will still be able to detect this. Certain kinds of lizards, insects also use this magnetic compass. Crustaceans or insects and mammals. A certain kind of cryptochrome used for magnetic navigation in flies has been detected in the human eyes. There is still debate as to whether it was used by humans for navigation as well.

Quantum Physics As Seen In Everyday Objects

At a point, we must feel irritated and slightly confused as to how so many concepts been mentioned here can be applied to our daily lives or used by the different instruments around us. Quantum Physics is one of the highlights of human intellectualism, and its knowledge has helped shape our civilization. Despite this relevance, most people still feel the subject of this field is quite abstruse and cannot be easily grasped by the ordinary mind. In the mind of the public, the concept of quantum physics is seen as a hard concept that is only understood by minds like Einstein and Hawking and another superhuman brain.

The concept of quantum physics is an understanding of the universe, and the universe is all around us, and its operation is based on the quantum rules. Even though we are so used to the laws of classical physics, and this relates to the universe at a macroscopic level, the understanding of quantum physics still affects various familiar operations. You'll find this list contains various tools and equipment that apply to quantum principle, without we realizing it.

TOASTERS

We are all familiar with the red glow produced by the heating element as we toast our bread. Funny enough, it was the observation if this red light that led physicists to ask questions, questions that birthed the quantum concept. Physicists wanted to know why hot objects shone that particular color of red, a very tough question, and quantum physics came to provide light to it.

Max Planck answered this issue in his theory, where he said that the light been transmitted must be discharged in discrete pieces of vitality, actual products of short, consistent occasions the recurrence of the light. For high-frequency light, the energy quantum is greater than the share of heat energy, which is assigned to that frequency, and this makes it impossible for light to be emitted at such frequency. This prevents the emission of high-frequency light. We could say that the toaster could be a central place where the idea of quantum physics first originated.

FLUORESCENT LIGHTS

The traditional incandescent light bulb was able to emit light by heating a piece of wire adequately until it gets hot and emit a bright white glow; this is similar to the phenomenon of the toaster. You are enjoying a groundbreaking work of quantum physics whenever you flick on a fluorescent bulb or one of the more recent twisty CFL bulbs; that's quantum physics at work.

In the early 19th century, physicists discovered that all elements found in the periodic table have a unique spectrum. When we heat a vapor of atoms, they will eventually emit light at a small number of discrete wavelengths, and each of the different items will have a different

pattern. The spectral lines were used to classify the composition of new material, and unknown elements such as helium were first discovered through this process.

This is how a fluorescent bulb works: whether the bulb is CFL or long tube, inside the bulb is a tiny bit of mercury vapor that is excited into plasma. Mercury easily emits light at frequencies that fall at the visible spectrum, and eyes will perceive this as white light.

CONCLUSION

Since its discovery, quantum physics has always been an emblematic subject, impossible to understand, probably because of its complicated language.
Yet in the last years it seems to be possible to register a greater interest of students (not only physics students) towards quantum physics, as the idea that the principles on which this incredible science is based explain the functioning of technological objects that we handle every day.

I would like to make an important clarification. Quantum Mechanics was born in the same years as the theory of relativity and has been, in a similar way, a leading theory throughout the 20th century. However, it has never managed to be a common subject for most people. One might speculate that this is due to the mathematical difficulties of the expressions governing the wave function and not just the complex plane and such. No, that's not enough to explain its "ghettoization".

But there is something else that seems to block its revelation. Relativity is not like that, and yet it has forcefully entered the everyday language. Moreover, Quantum Mechanics is at the premise of today's apparent multitude of mechanical developments, from nuclear vitality to Computer, electronic devices, from advanced tickers to lasers, semiconductor frameworks, photoelectric cells, and indicative and treatment hardware for certain diseases.

To put it bluntly, we can "live" in a "cutting edge" way today, thanks to quantum mechanics and its applications.

Our brain, seems to be based on quantum processes, including state superpositions, wave collapses and entanglement situations. The complication lies in its "counterintuitive" postulates about the reality of nature. The difficulty is to enter in an unknown and absurd world as Alice in Wonderland. This does not mean that we should have inferiority complexes about these concepts. The founding fathers themselves experienced this situation at the limit of absurdity. Could it be that nature followed completely arbitrary rules or, was it all an appearance caused by the lack of information, of deterministic type, still missing? The same creator of the general and ultra-confirmed uncertainty principle (Heisenberg) said: "I remember the long discussions with Bohr, which made us stay up late at night and left us in a state of deep depression, not to say of true despair. I kept walking alone in the park and kept thinking that it was impossible for nature to be as absurd as it appeared to us from the experiments." Basically, there is no defined and describable reality, but an objectively indistinct reality composed of overlapping states.

Let us take a couple of fundamental concepts that we have come to know, but certainly not understand:

1.   Every action of the finest structure of matter is characterized solely and only by its probability of occurrence. Fully causal, non-deterministic phenomena. But, above all, by the indistinct separation between the observed object, the measuring instrument, and the observer.

2.   Surely, under certain conditions, what happens in a certain place can drastically influence what happens in a completely different place, instantaneously. This explains the phenomenon of entanglement, the interweaving of particles that have had an interaction in their past (but recent research seems to admit also "contacts" in the future) or that were born "together". Although separate, they always represent the same entity. An action taken on one has an instantaneous effect on the other.

I guess you have already understood what the real problems of quantum physics are. They arise both from the difficulty of dealing with concepts too far from everyday reality, and the difficulty of using an adequate language to explain this absurd world. Mathematics can also describe it, but the letters and the words of this strange alphabet are missing. Exceptional, in this sense, was the work of Feynman with his diagrams applied to QED (which we now know quite well).

I could not help quoting a phrase by Max Born in this regard: "The definitive beginning of the problem lies in the reality (or philosophical rule) that we are forced to use the expressions of basic language when we want to represent a marvel, not with a legitimate or numerical examination, but with an image that speaks to the creative mind. Everyday language has evolved from everyday experience and can never overcome these limiting points. Traditional material science has limited itself to the use of such ideas; by examining perceptible movements, it has created two different ways of speaking to them by basic cycles: moving particles and waves. There is no other method of giving a pictorial representation of movements - we must apply it also in the area of nuclear cycles, where old-style material science separates."

More or less, the representation of QM itself could be intensely influenced by our "artwork" graphical cut-off points.

Thus, the founding fathers themselves often used analogies and similes to express purely mathematical concepts. They must, however, be taken for what they are and not given any real, concrete validity.

This is a huge problem for our brain (especially today) even if - maybe - it would have all the bases to use an adequate language, but still too indistinct to be formulated correctly: Feynman's diagrams, I repeat, are a wonderful attempt in that direction.

Niels Bohr himself used graphic analogies to try to support theories so absurd for our classical language. Famous is the white vase that represents, at the same time, two black social profiles.

A state of superposition between two realities existing instantaneously (two states or - perhaps - two universes?). This type of analogy has influenced many optical illusion games and even artistic currents (think Picasso).

It is a pity that these interpretative efforts, together with the more complete and refined ones of Feynman, do not find space in schools to adequately prepare young people to "stutter" their first quantum words and to start a primitive language that would allow them, today, to understand, at least partially, the reality of Alice. And not just passively undergo the most wonderful technological applications that are now an integral part of their physical bodies. Real "appendages", which however act unconsciously, independently from any mental command.

Unconditional reflexes and nothing more. Moreover, de Broglie advanced his bold hypothesis just following the symmetries of visible Nature. He associated only to matter in general, what happened to light.

So basically: if light manifests itself under a double aspect, undulatory and corpuscular, why not think that matter follows the same rule? It is enough to associate to each corpuscle of matter a wave of a certain length, that is a phenomenon extended to the space surrounding the particle. The dualistic nature (particle-wave) applies to all particles, such as electrons, atoms and other moving entities. However, the basic problem (still subject to discussion and interpretation), which we have mentioned, remains open. The matter wave that commands the particle can be deterministic, and therefore still unknown in its real structure (in line with Einstein's idea) or, instead, a different representation of the same particle and therefore follow the rules of a complete causality (Copenhagen school).

One way or another, however, it must be concluded that light or a beam of electrons is nothing more than a "train" of electromagnetic waves, but also a jet of "bullets" as in the double slit experiment.

While remaining in this basic ambiguity, Schrödinger formulated the equation that perfectly describes every adulatory property of matter through its wave function. It allows us to describe each and every behavior and, more importantly, to calculate the probability distribution of finding a particle within the associated wave.

Overwhelming mathematics that, however, does not negate the fact that Schrödinger himself did not believe in the actual concreteness of this representation. Everything and the opposite of everything (conceptually), but described in the same way.
However, his equation confirms what Feynman's experiment illustrated above: a particle can occupy ALL possible positions within the associated wave. By occupying all possible positions, it no longer has an actual place of existence or direction. It automatically nullifies any possible prediction of its future except in purely probabilistic terms (QED is always more understandable... don't you think?). The pilot wave or a hidden variable does not change Nature's action and its probabilistic description.

Once again, we fall back on Heisenberg's principle. In classical mechanics, the deterministic essence automatically allows one to predict the future if one has exact information about the position and velocity of a particle. We remember, in this regard, that the first mathematical methods that allowed the calculation of an orbit of a "planetary" particle were based (and are still based) on the knowledge of at least three positions and three speeds, such as to allow the solution of an orbit characterized by six unknowns. Too easy for microscopic particles.

The probabilistic conception leads inexorably to the uncertainty principle, inherent to all microcosm: either you know the position, or you know the speed. To know both with precision is impossible.

Otherwise, the particle would be localized and the wave would collapse. And we go back again to the starting point. If there is an initial causality (completely unknown) or if there is no causality at all. In a nutshell, the double-slit experiment perfectly illustrates all the problems of Quantum Mechanics.

It is worth reflecting on Einstein's dramatic emotional situation. While he was giving physical reality a perfectly deterministic representation, he found himself involved in a representation that led to the complete causality of nature. He said," The theories of quantum radiation interest me greatly, but I would not want to be forced to abandon strict causality without trying to defend it to the limit.

Yet, no physicist has contributed as much as Einstein to the creation of quantum physics. What he demonstrated about it (and for it) was sufficient and advanced a scientific career of the highest order (not for nothing did it earn him the Nobel Prize). It is therefore easy to understand his existential drama, which never abandoned him until his death. A mixture of anger, wounded pride, unshakable confidence, and despair for not being able to prove his certainties. This mix of frustration, exaltation, hope, disappointment, innovation, and conservatism permeated all the great minds that gave birth to the

Quantum Mechanics. A very choral work and certainly not a puzzle of individual ideas. Everyone, almost unwillingly (sometimes even against their own purposes), did nothing but put one more brick to a building that was becoming an incredible skyscraper with increasingly solid and unassailable foundations.

Maybe this unique way in the history of science to formulate a more and more complete and refined theory, by many superior minds, could make us understand that QM is something inherent in the human mind, but that has extreme difficulty to come out. Surely the knowledge of the language of Classical Physics has made great strides, but not so far from the almost unconscious intuitions of Democritus and Epicurus.
Simply put, the mind must be trained to go along with a reality that is only historically and culturally absurd. The more we go into the very essence of quantum physics and its principles, the more the double slit experiment becomes fundamental and complete. A true scientific masterpiece, a manifesto itself of the future of human intellect.

QM will reveal us the future.